Patrick Kamalenga Ngoyi

Introduction to the study of light dispersion

Patrick Kamalenga Ngoyi

Introduction to the study of light dispersion

Geometric Optics

ScienciaScripts

Imprint

Cover image: www.ingimage.com

This book is a translation from the original published under ISBN 978-620-6-71367-8.

Publisher:
Sciencia Scripts
is a trademark of
Dodo Books Indian Ocean Ltd. and OmniScriptum S.R.L publishing group

120 High Road, East Finchley, London, N2 9ED, United Kingdom
Str. Armeneasca 28/1, office 1, Chisinau MD-2012, Republic of Moldova, Europe
Printed at: see last page
ISBN: 978-620-7-98130-4

Contents

The aim of physics is to understand the mechanisms that govern the behaviour of the natural and artificial things in our "environment".

The development of optics involves a wide range of approaches.

Mathematics made a major contribution to the development of optics, as in the case of rectilinear trigonometry, which is widely used in this branch of physics.

Optics is not just a matter for physicists, but for all scientists and all those who know a little or a lot, but who are duty bound to share their knowledge with others.

That being the case, we are asking ourselves the following questions, which also form our problem.

a) What is light?
b) How does light break down?
c) What are light-dispersive devices?

Light is omnipresent in our lives, and it is thanks to light that life is possible on our planet. Life could not have developed without sunlight.

Thus, ordinary white light in its state decomposed by the prism would be made up of 3 fundamental colours (primary colours) seen by the eye.

This work is intended for physics learners, teachers and scientific researchers.

In addition to the general introduction and conclusion, this work is divided into three chapters:

- The first chapter is a general introduction to geometrical optics;
- The second chapter deals with light dispersion.
- The third chapter deals with the applications of light scattering.

CHAPTER I

GENERALITIES ON GEOMETRICAL OPTICS

1.1.OBJECT OF GEOMETRICAL OPTICS

Geometrical optics studies the formation of images by optical systems.

1.1.1. Optical systems

A. Definition

An optical system is any element capable of modifying the propagation of rays from a point on an object. An optical system is made up of transparent, homogeneous and isotropic media.

B. Images

An image is defined in relation to a given optical system. It is the set of points where the light rays converge.

C. Virtual objects and images

An object point A emits light rays towards the optical system. There are two possible cases:

1) The rays emerging from the optical system converge on a point д: this point is a real image point and can be collected on a screen.

2) The rays emerge from the optical system diverging, but their extension intersects at a point A': this point is a virtual image point, it cannot be seen 3) r on the screen, but it cannot be seen by the naked eye through the system.

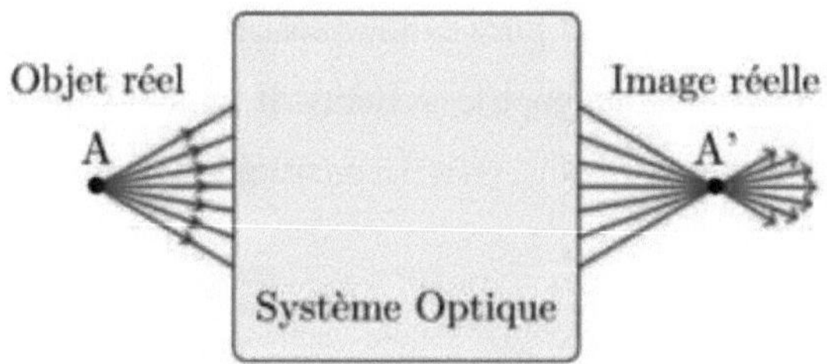

D. Homes

The foci of an optical system are particular points defined as follows:

1) The principal image focus F' is the image point of an object located at infinity whose rays arrive parallel on the optical system and parallel to its optical axis. The plane passing through F' and perpendicular to the optical axis of the system is called the image focal plane.

2) The principal object focus A is the object point of an image located at infinity, with the rays emerging from the optical system parallel to each other and parallel to the optical axis. The plane passing through F is perpendicular to the optical axis of the system and is called the object focal plane.

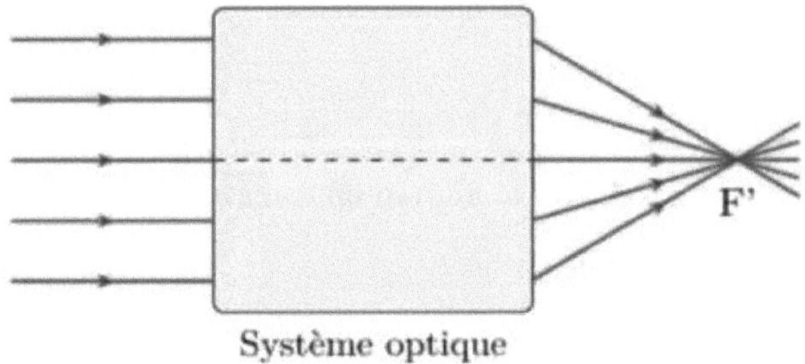

E. Goal

Optics studies luminous phenomena, i.e. mainly phenomena perceived by the eye. The cause of these phenomena is light, because in order to be visible, an object must transmit light to the eye.

Geometrical optics is a branch of science based on the notion of the light beam. This simple approach allows for the geometric construction of images, which gives it its name. Geometrical optics is the most flexible and effective tool for dealing with dioptric and catadioptric systems. It explains the formation of the images produced by these systems. **(www.consulté le 6 juillet à 20h22min)**

1.2.THE LIGHT

1.2.1. Notions

Light is omnipresent in our lives, and it is thanks to light that life is possible on our planet. Life could not have developed without sunlight. Even today, plants and animals need light to survive.

Light is also our main means of discovering the world around us. It is estimated that the vast majority of information received by our brain about our environment is provided by our eyes. The eye is in fact the most sophisticated optical instrument we know.

Over the centuries, people have discovered the properties of light and have gone on to design numerous instruments that make use of these properties.

Some, like the telescope or the microscope, have enabled us to discover previously unknown worlds. Others, such as eyeglasses or surgical lasers, have improved our quality of life **(KAMALENGA NGOYI PATRICK,** *Classical studies of optical instruments*, **TFC, 2014-2015, ISP/MBM).**

1.2.2. What is light?

Light is a form of energy, just like electricity or heat. It is made up of tiny particles called photons and travels in the form of a wave.

Light is actually generated by the vibrations of electrons in atoms. It is therefore a mixture of electric and magnetic waves: light is said to be an electromagnetic wave.

There are several forms of light. The one we know is visible light. However, there are several other forms of light waves: infrared, ultraviolet, X-rays, etc.

The difference between these types of light is their wavelength or the amount of energy they carry. For example, a wave can be created by attaching a long string to a wall. By waving the other end up and down, we create ripples that propagate through the rope.

The distance between two neighbouring waves is the wavelength (Á). The longer the wavelength of light, the less energy the photon has.

Visible light has the longest wavelength but the least energy. In contrast, blue light has the shortest wavelength and the most energy.

The different forms of light are classified in what is known as the electromagnetic spectrum. Visible light occupies only a very small portion of this spectrum. At slightly longer wavelengths, we find infrared light, which gives us the sensation of heat... Then there are the microwaves used in ovens and radars.

Finally, radio waves, which carry radio and television signals, have the longest wavelengths: up to a few metres.

Beyond visible light, towards the shorter wavelength regions, we find ultraviolet light, which makes us tan, and X-rays. This type of light has the property of passing through the soft tissues of the human body and being absorbed by bones and teeth. X-rays are used, among other things, to produce X-rays. Gamma rays are the photons with the shortest wavelengths and the highest energy.

In a vacuum, light travels in a straight line at a speed of almost 300,000 km/s. At this speed, we could go around the Earth seven and a half times in one second. This is the universal speed limit. Nothing in the universe can go faster than light.

1.2.3. Wave nature

(According to Prof GEORGES MWENDA KAZADI, *lecture note on special questions in* **undergraduate** *physics*)**,** light is the energy of electric and magnetic nature in motion.

This energy is transported from one link to another in the form of waves.

Light can propagate in a vacuum (we can see the stars) and in air as well as in matter in the form of straight rays of light, where its celerity is maximum, equal to $C=3.10^8$ m/s. It is a transverse vibration, perpendicular to the direction of propagation, with a frequency of the order of 10^{15} Hz.

The light wave can correspond to a sinusoidal vibration of given frequency in the electromagnetic field. It is then called a monochromatic wave because the sensation of colour depends on the frequency. Monochromatic radiation is characterised by its frequency (f, N) and wavelength.

$$\lambda = \frac{c}{f} \qquad (1,1)$$

Where λ is the wavelength in metres

***C** is* the speed of light in a vacuum in metres per second.

F is the frequency in Hertz.

Light can also correspond to a non-sinusoidal vibration, whether periodic or not, for which a frequency spectrum can be defined, either wheel-shaped or continuous.

We can therefore consider light as a wave characterised by: its wavelength X, its frequency (f, N), its celerity (c) and its amplitude.

There is no exchange of energy (the medium is not dissipative), only preparation, interference and diffraction. Its wavelength is infinitesimally small.

In addition to the wave aspect **(prof. Georges KABAMBA MWENDA KAZADI, questions spéciales de physique L1 math-physique, ISP/MJM),** electromagnetic waves can be represented under a corpuscular aspect, as formed of photons or quantum electromagnetic radiation, elementary units of zero rest mass, whose energy (E, N), representing an elementary quantum, is proportional to the frequency (f, N) and given by Planck's law.

$$E = H.f \qquad (1,2)$$

Where

E is Energy in joules,

H is Planck's universal constant in joules per second ($h=6.624.10^{-34}$ J.s)

f is the frequency in hertz

Energy is exchanged, emitted, absorbed, shocked...

The range of electromagnetic wavelengths perceptible to the human eye is between 400 and 700nm. Wavelengths shorter than 350nm (UV) or longer than 1400nm (IR) are absorbed in the 'transparent' media of the eye, so only visible light deserves this label, as the near ultraviolet is part of the infrared.

Under certain circumstances, light acts as if it were made up of corpuscles. Thus the photoelectric effect, one application of which is the magic eye (photoelectric cell) that opens doors for you in certain department stores.

1.2.4. Light propagation

Light travels in a straight line.

Its path is represented by straight lines with an arrow indicating the direction of travel (from the source). The path of light is not visible in a vacuum or in air.

A. Primary light source

It's an object that creates its own light.

Examples: the sun, the stars, a lighted lamp, etc.

B. Secondary light source or scattering object

It is an object that reflects the light it receives in all directions.

Examples: the moon, the planets, etc.

In general, all bodies that emit light are light sources. These sources can be :

1) Transparent

That lets the light through.

2) Opaque

That doesn't let the light through

3) Translucent

Allows light to pass through, but does not allow objects to be perceived accurately (blurred vision). **A. de Laruelle, A. I. Claes, cours de physique, chaleur, énergie, optique géométrique. Wesmaiel Charlier (S.A) Namir 1971.**

C. Light beam

A light ray is the straight line along which light propagates in a homogeneous medium.

D. Light beams

Light beams are an infinite group of light rays.

E. Types of light beams

1) Parallel or cylindrical beam

A light beam is said to be parallel or cylindrical when all its components are parallel.

Example

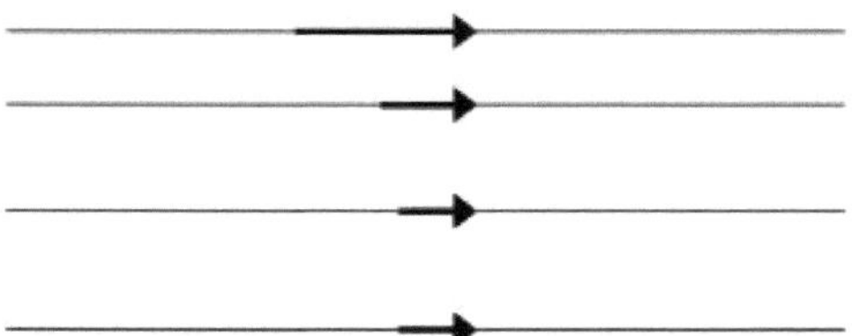

2) The divergent beam

A beam of light is said to be divergent when all its component rays come from the same point.

3) The convergent beam

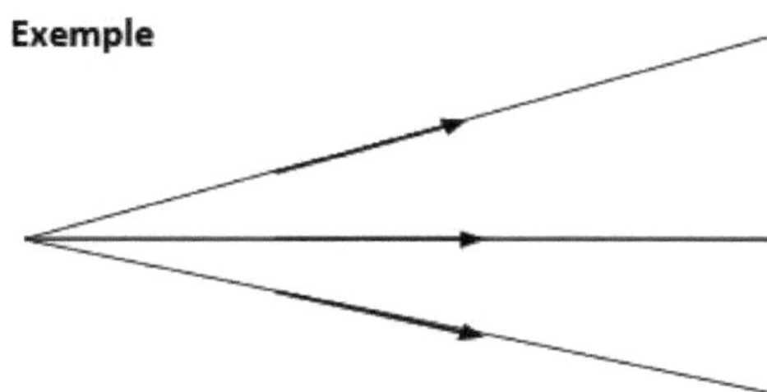

A beam of light is said to converge when all its component rays are directed towards the same point.

Example

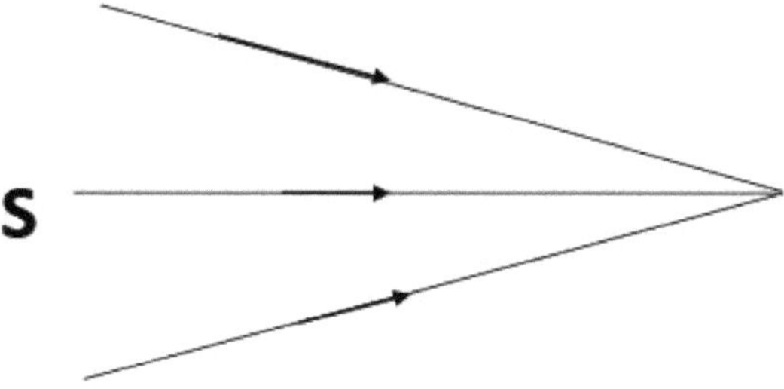

Note

A very narrow beam is called a "light brush". This is the case, for example, of the beam that goes from a point of light to the eye.

1.3.REFLECTION OF LIGHT

1.3.1. Definition

The reflection of light is a sudden change of direction that light rays undergo when they hit a reflective surface (mirror, thin layer of water, etc.).

1.3.2. Types of reflection

There are two types of light reflection:

- ✓ Regular reflection
- ✓ Diffused reflection.

A. Regular reflection

1) Definition: The reflection of light is said to be regular when a beam of parallel rays falling on a reflecting surface changes direction so that the reflected rays are parallel.

2) Example

B. Diffuse reflection

1) Definition

Light is said to be diffusely reflected when it is reflected in all directions, i.e. the reflected rays are dispersed and do not follow any specific direction.

This reflection occurs on irregular (or unpolished) surfaces. Light is reflected in several directions.

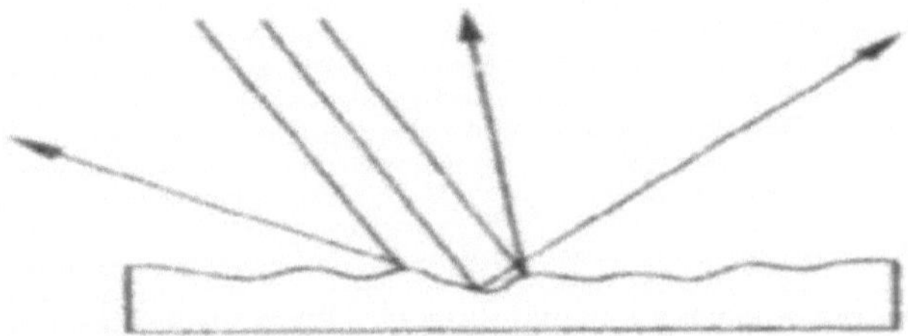

1.3.3. Laws of reflection

Reflection occurs when a ray of light suddenly changes direction while remaining in the same propagation medium.

Experience shows the following laws of reflection:

a) The incident ray, the reflected ray and the normal to the surface are in the same plane, called the plane of incidence.

b) The angles of incidence and reflection are equal.

N.B.: the direction of the angles is arbitrary, so we choose a positive direction and stick with it. However, the angles are always oriented from the normal.

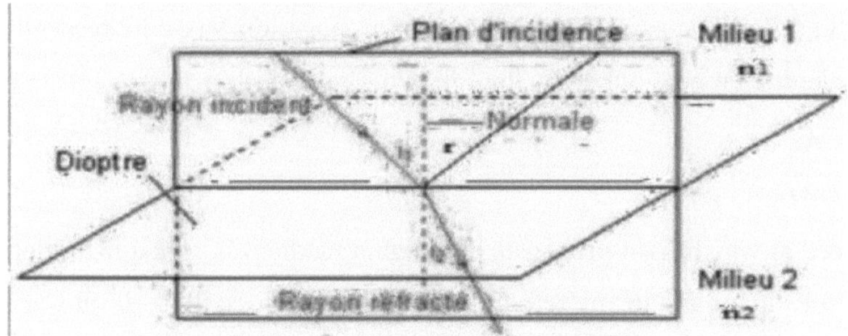

1.4. REFRACTION OF LIGHT

1.4.1. Definition

The refraction of light is a sudden change of direction that a ray of light undergoes when crossing the separating surface of two transparent media.

The surface separating two transparent media is called a "dioptre".

The ray propagating in the first medium is called the "incident ray" and the ray propagating in the second medium is called the "refracted ray".

Experience shows that refraction obeys the following laws:

a) **1ère law:** the incident ray, the refracted ray and the normal to the surface are in the same plane of incidence.

b) **2nd law**: the angles of incidence *Γ* and refraction *f* are linked by **SNELL DESCARTES**' law**:**

$$n = \frac{\sin \hat{\imath}}{\sin \hat{r}} \qquad (1,3)$$

Where = refractive index of the first medium relative to the second.

1.4.2. Remarks

> If n is greater than 1 (n>1), the second medium is more refractive than the first. This is the case when light passes from air into glass; into water; etc.

> If n 1, the second medium is less refractive than the first, the ray deviates from the normal to the dioptre. So the angle , <f ·

1.4.3. Refractive index and speed of light

The phenomena of refraction is due to the fact that light propagates at different speeds in different media. The velocities of propagation Vi *e t V2* respectively in the media; the í n 2 are linked by the following relationship:

$$\frac{v_1}{v_2} = \frac{\sin \hat{i}}{\sin \hat{r}} = n \qquad (1,4)$$

The speed at which light propagates through a medium is called its "celerity". It is 3.10^8 m/s in air and in any refractive medium; it is less: $2.25.10^8$ m/s and 2.10^8 m/s in glass.

1.4.4. Refraction limit angle

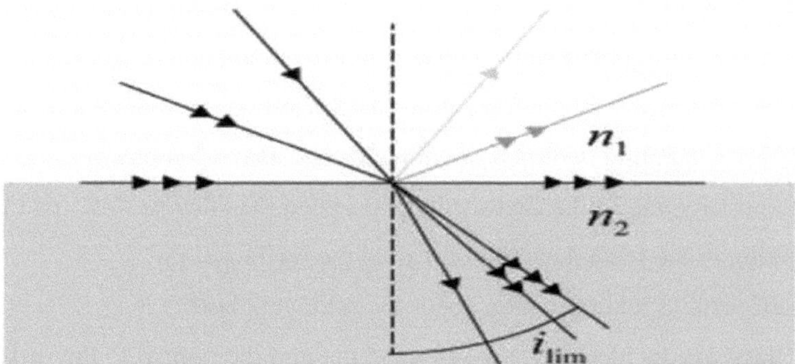

Consider the passage of air through water. The deviation increases as the index becomes greater obliquely at an incidence of 90° when the incident ray grazes. The angle of refraction reaches its maximum value, the angle of refraction limit.

$$n \sin \hat{r} = \sin 90°$$

$$\sin \hat{r} = \frac{\sin 90°}{n}$$

$$\sin \hat{r} = \frac{1}{n} \qquad (1,5)$$

1.4.5. Discussion of the laws of snell-descartes

> The angle of incidence ¿can take any value between 0° and 90°;

> The first law allows us to draw the same conclusion for the angle of refraction **r** ;

^ 11-

> The values of the refracted angle are related to those of $\hat{\imath}\ et\ \frac{n_2}{n_i}$

> Propagation towards a more refractive medium refraction limit. The refracted bray$_i$ ·<$_i$ exists and reaches a limit value given by $\hat{\imath} = 90°\ ;\ \sin\hat{r} = \frac{n_1}{n_2}$

In this case, we speak of limiting refraction.

> Propagation towards a less refractive medium: total reflection.

The refracted ray $\hat{r} > \hat{\imath}$ no longer exists for an incidence greater than a limit value set by $\sin \hat{\imath} = \frac{n_2}{n_1}$. In this case, we speak of total reflection.

> If the angles are small, *sin r,* Snell-descartes' law takes the form

Kepler's Law $n_1\hat{\imath} = n_2\hat{r}$ **Planar diopters**

A Blades with parallel faces

1) Definition

A transparent medium bounded by two plane-parallel faces (two plane-parallel diopters) is called a parallel-face lens.

B. The path of a light beam passing through a plate with parallel faces

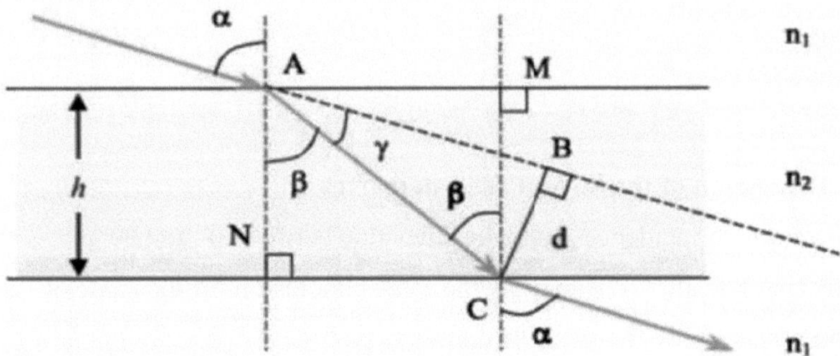

D = parallel displacement = lateral displacement

Law of refraction in A :

$$n_1 . \sin\alpha = n_2 . \sin\beta \qquad (1,6)$$

Because of the symmetry of the problem, we have the same angles at C as at A, so here we also obtain the law of refraction:

$$n_2 . \sin\beta = n_1 . \sin\alpha \qquad (1,7)$$

The light beam leaving the plate is parallel to the incident light beam **(CT Médard NSUNGULA,** *physics II math-physics*, **ISP/MJM).**

CONCLUSION

✓ Lateral displacement increases with incidence.

✓ For the same angle of incidence, lateral displacement increases with thickness. So light is a physical agent capable of making an impression on the retina of the eye.

CHAPTER II

LIGHT DISPERSION

2.1.DEFINITION

Light dispersion is a phenomenon in which light breaks down into a spectrum of coloured light ranging from violet to red.

2.2.LIGHT DISPERSIVE DEVICES

These are optical systems that cause light to decompose.

2.2.1. Prisms

A. Definition

A prism is a transparent medium separated by two non-parallel faces. A ray of light passing through it undergoes double refraction

B. Description of the prism

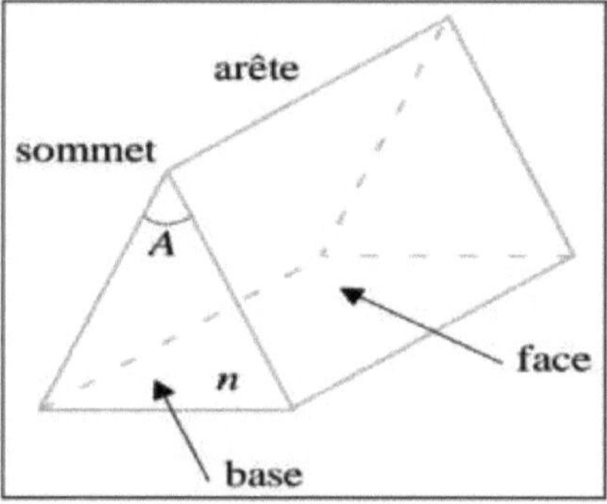

A prism is made up of a transparent medium bounded by two flat faces. It is characterised by its vertex angle A and its refractive index n.

C. Action of a prism on a ray of light

A ray of light striking one of the faces of the prism is refracted, then passes through the medium of index n and is refracted again (see figure 2.2).

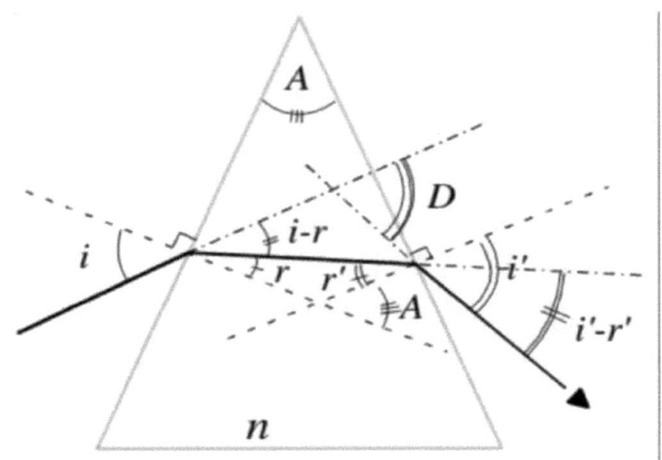

For the first refraction, the angle of incidence is denoted by *i and* the angle of refraction is denoted by *r.* For the second refraction, the angle of incidence is denoted by ?\ and the angle of refraction *is* denoted *by* ^ı-.

D. How a prism works

The principle of an optical prism is based on the principle of refraction of light. When light passes from the air to the glass of an optical prism, the beam of light is refracted, which also happens when it emerges from the other side. This emerging ray is parallel to the incident ray.

Because the system contains a reflection, the output image is inverted. Top becomes bottom or right becomes left.

To use the prism correctly, the incident ray must lie in a plane perpendicular to the edge of the prism (if not, everything happens as if the angle at the top of the prism were variable).

Under the correct conditions, the plane of incidence and refraction all coincide with the cross-section of the prism.

E. Effect of premium

The effect of the prism is therefore to **deflect the light beam.** The deflection, which is the angle between the incident (initial) ray and the emergent (final refracted) ray, is denoted D. **(www.google.com consulted on 25/5/2023 at**

19h51min)

F. Properties of a prism

A prism is a 'special' triangular-shaped lens (formed by two non-parallel faces that form an angle with each other) that has the property of **deflecting** or **redirecting** the image of the object towards its apex, making it 'fall' correctly onto the retina of each eye.

G. Prism characteristics

The prism has four characteristics:

a. First, the laws of refraction:

$$\boldsymbol{sin\hat{\imath} = nsin\hat{r}} \qquad (2,1)$$

$$\boldsymbol{nsin\hat{r}' = nsin\hat{\imath}'} \qquad (2,2)$$

The laws governing plans have already been used to draw the diagram.

b. Then a geometric constraint linking A,,i ^, and r^v . The angle between the normals to the faces of the prism is equal to A (see figure 2.2) and the sum of the angles of a triangle is equal to 180° :

$$\hat{r} + \hat{r}' + (180°\text{-}A) = 180° \qquad (2,3)$$

$$\hat{r} + \hat{r}' = A$$

Finally, the expression for the deviation :

$$D = \hat{\imath} - \hat{r} + \hat{\imath}' - \hat{r}' = \hat{\imath} + \hat{\imath}' - (\hat{r} + \hat{r}') = \hat{\imath} + \hat{\imath}' - A$$

$$D = \hat{\imath} + \hat{\imath}' - A \qquad (2,4)$$

H. Influence of refractive index on deflection

The deviation is given by $D = t \mid r$,↑ and we are trying to find out how to deviate D when the index n varies. (As glass is dispersive, its index varies with wavelength).

For a given incident ray, the angle of incidence Γ is fixed and A is a constant parameter for a given prism, so D depends on Γ. We therefore need to find out how Γ 'varies as a function of n. Let's proceed in stages:

During the first refraction r is given by . $sin\hat{r} = \frac{sin\,\hat{\imath}}{n}$; so when n increases. Then r^{v} is given by$_{; = t}$, ; so f increases as r decreases.

Finally, the second refraction given by sin $\ddot{\imath}$ '= nsinf , so f increases when e and n increase, so r increases when n increases.

I. Influence of wavelength on deflection

To a first approximation, Cauchy's law gives variations in the refractive index of a glass as a function of wavelength:

b = T (a-n)

Where a and b are constants characterising the transparent medium.

This law shows that the index increases as the wavelength decreases, i.e. as we go from red to violet.

As a result, red light will be less deflected than violet light. This is in line with experimental findings. **(A. DELARUELLE and AI. CLAES, éléments de physique, chaleur-acoustique-optique, NAMUR, pages 1966).**

J. Emergence condition

There are two emergence conditions, one relating to the angle of incidence and the other to the angle of the prism. **1)**

1) Incidence angle condition

For the emergent ray to exist, it is necessary that $f \leq fUm$, *the* limit angle of incidence, because when / *rUm* there is total reflection on the second face of the prism. When f$\leq$ *Γιim* the angle *i'* is 90°, so the angle f is given by :

$$n sin\hat{r}' lim = \sin 90° = 1$$

$$sin\hat{r}' lim = \frac{1}{n} \qquad (2,6)$$

The condition > _ ∏ι_{in} imposes$_{f}$ - ≥ л fu_m because$_7$ ^ = л - f from relation (2.5) and this imposes .$_s$ ⁄m > H.WI(⁄1 Λ Iim) from relation (2.1), so: i ≥ ⅛

With :

$$sin\hat{\imath}_0 = nsin(A - \hat{r}' \lim) \qquad (2,7)$$

2) Prism angle condition

Furthermore, the angle of refraction f is always less than the refraction limit angle f *11 m, which is* equal to the incidence limit angle *f '1 ï.* Consequently, the two inequalities: f' ≤*f'ιim* and *f* ≤ r/rm are verified simultaneously, or f + f ' - л , so :

$$A \leq 2\hat{r}\ \mathrm{lim}\ avec\ sin\hat{r}\,lim = \frac{1}{n} \qquad (2,8)$$

When A≤ *2flιm,* there is always total reflection.

K. Total reflection prism

This is a prism whose cross-section is an isosceles right-angled triangle.

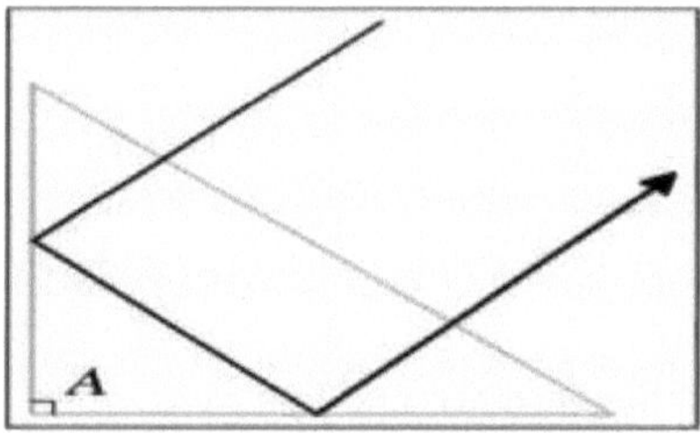

The light beam is directed towards the base of the prism at normal incidence. The ray penetrates the prism without deflection and then strikes the first face at an angle of incidence 45° greater than the limit angle of incidence; there is therefore total reflection. The second face acts in the same way because the angle of incidence is again 45°.

This allows the prism to be used in a different way, i.e. to 'bend' the light beam. This type of prism is used in binoculars and terrestrial scopes to reduce clutter and straighten the image (otherwise, the sky would be at the bottom and the ground at the top). This requires two total reflection prisms with orthogonal edges

(**KAMALENGA NGOYI**, *Classical studies of optical instruments*. Cas de l'appareil photographique, lanterne de projection et lunette, TFC edition 20142015, ISP/MBM).

L. Minimum prism deflection

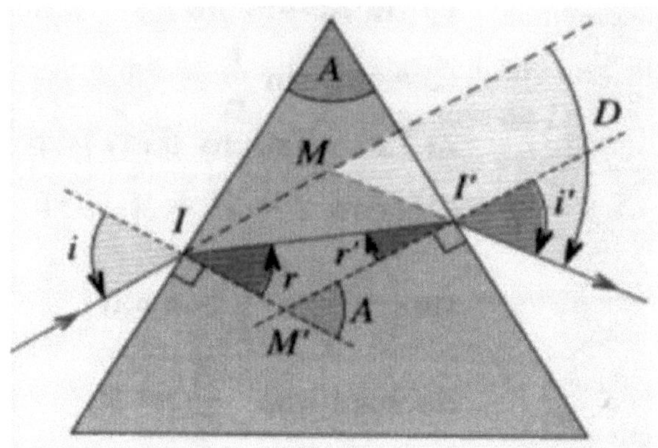

The angle of deflection D is D= ı - f + r - *f* or ^ = *f* + *f* ' where A is the angle at the apex of the prism.

Hence D = i + i-.4

D is minimum when $\frac{dD}{Di} = 0 \text{ soit } r + \frac{di\prime}{di} = 0 \text{ et donc } d\hat{\imath}' = -\,d\hat{\imath}$

The laws of refraction at the entrance and exit give

$$\sin(\hat{\imath}) = n s\hat{\imath}(\hat{r}) \qquad (2,9)$$

$$\sin(\hat{\imath}') = n s\hat{\imath}(\hat{r}') \qquad (2,10)$$

By differentiating (2.9) and (2.10), we obtain cos(i) *di* = *ncos(f) df* and cos(√) *di'* = *ncos(*f ') *df:*

$$\frac{\cos(i)\,di}{\cos(i')\,di'} = \frac{n\cos(r)dr}{n\cos(r')dr'} \qquad (2,11)$$

We have shown that *di* = *-di'* and by differentiating *Á* = *f'* + f ', we also obtain that

d E -d,■. Replacing in (2,9), we then have : $\frac{\cos(i)}{\cos(i')} = \frac{\cos(r)}{\cos(r')}$

We can then easily obtain $\sin(\hat{\imath})^2$

Or

$$\frac{1-\sin(i)^2}{n^2 sin(i)^2} = \frac{1-\sin(i')^2}{n^2-\sin(i)^2} \qquad (2,12)$$

To solve this equation, we need to study the function

$$f(x) = \frac{1-x}{n^2-x} \text{ pour } x > 0$$

Let's go

$$f(x) = \frac{-1}{n^2-x} + \frac{1-x}{(n^2-x)^2} = \frac{1-n^2}{(n^2-x)^2} < 0 \; car \; n > 1$$

The function *f* is monotone decreasing and therefore injective. Equation (2.12) therefore has the unique solution □ˆ= □ˆ', which immediately implies that $\hat{r} = \hat{r}'$.

So $Dm = 2i - A = 2\arcsin\left(nsin\left(\frac{A}{2}\right)\right) - A$

Or $\frac{A+D}{2} = \arcsin(nsin\left(\frac{A}{2}\right))$

From this we can derive the classic formula giving the index of the prism's glass as a function of

of *Dm :*

$$n = \frac{\sin(\frac{A+Dm}{2})}{A} \text{ Sin(2)} \qquad (2,13)$$

M. Graphical representation

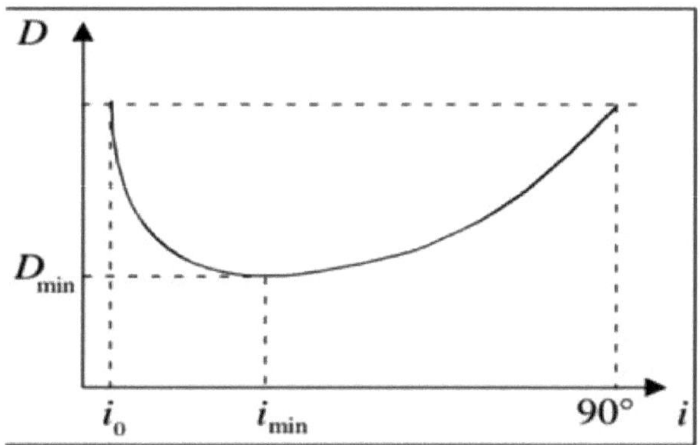

2.2.2. Network

A. Definition

The grating is a very thin optical surface made up of a very large number of identical, equidistant fine slits.

B. Description of a network

A grating consists of a transparent support on which a large number of fine, parallel and equidistant lines have been engraved. Each line represents a slit that diffracts the light.

C. Network characteristics

The network is characterised by its "pitch *a*", which corresponds to the distance between two consecutive lines, and also by the number "*n*" of lines per unit length (around 500 lines per millimetre).

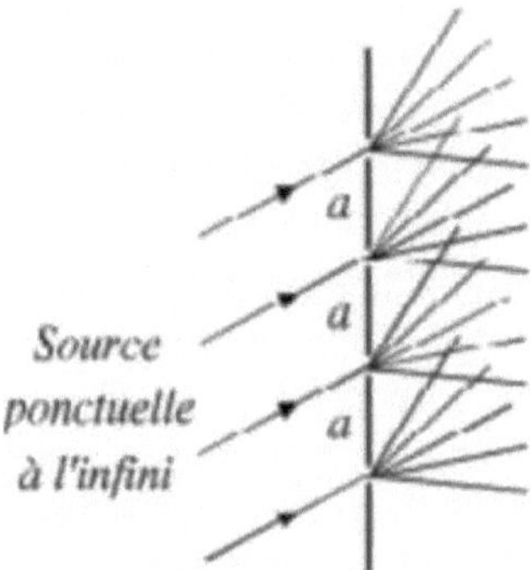

D. Relationship between *a* and *n* in the network

There is a relationship between the pitch *a* of the network and the number n of lines per unit length.

$$a = \frac{1}{n} \qquad (2,14)$$

E. Diffraction by a grating

A grating diffracts light of wavelength in several orders p. Thus, in transmission, a parallel beam under incidence *Γ* is deviated to order p in the direction* such that :

$sin\hat{r} = pn\,\lambda + sin\hat{i}$ (2,15)

(2,(15) **(J. DESSART, J. C. JODOGNE et JODOGNE, optique géométrique, édition A. DE BOECK-BRUSSELS).**

F. Feature density

The line density is defined as?î !(number of lines per metre). The table

CL

The following table gives orders of magnitude for different types of network:

Quality	(m 1)	n(lines per incl.)	A(um) width	N	
Average		103	30	2 cm	2000
Classic	4.105	10^4	3	3 cm	10000
Excellent	4.10	105	0,3	4 cm	40000

2.3. DISPERSION OF LIGHT BY THE PRISM

2.3.1. Dispersion of white light

A. White light

White light is made up of a set of different colours that make up the light spectrum.

This spectrum is polychromatic and continuous: it contains all the shades of colour between violet and red.

B. Creating white with the colours of the rainbow

The rainbow is a luminous phenomenon that can be seen in the sky.

When the sun shines through the rain, we see different colours spread out like a ribbon in the shape of a bow.

2.3.2. Newton's experiment

A prism is illuminated with a beam of white light.

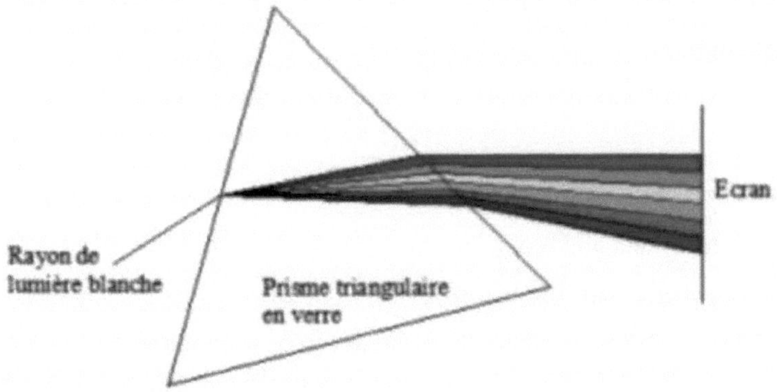

A. Comment

- ❖ When white light passes through the prism, we see a rainbow-like display of colour. The prism is said to decompose white light.
- ❖ The coloured pattern obtained is called the spectrum of white light.
- ❖ The seven main colours of the white light spectrum are: violet, indigo, blue, green, yellow, orange and red.

In optics, it is a physical phenomenon that causes the separation of a wave in the spectral component with different wavelengths, due to the dependence of the speed of the wave on the wavelength in the medium traversed. It is often described in light waves, but can occur in any type of wave that interacts with a medium or that can be confined in a waveguide, such as sound waves. Dispersion is also called chromatic dispersion to emphasise its dependence on wavelength. A medium that exhibits these characteristics against wave propagation is said to be dispersive.

B. Description

There are generally two sources of dispersion: **material dispersion**, which derives from the fact that the response of the material to waves depends on the frequency, and **waveguide dispersion**, which occurs when the speed of the wave in the guide depends on its frequency. Transverse wave paths in a confined waveguide space generally have different finite velocities (and field shapes), which depend on the frequency (for example, by the relative size of a wave, the wavelength, compared to the size of the guide).

Dispersion in waveguides used for telecommunications involves signal degradation, as the different delays with which the different spectral components arrive at the receiver 'dirty' the signal over time or create distortion. A similar phenomenon is modal dispersion, caused by the presence of several modes in a guide at a given frequency. A special case is polarisation mode dispersion or PMD (polarisation mode dispersion), which derives from the composition of two separate polarisation modes moving at different speeds due to Randomich imperfections that break the symmetry of the guide.

C. Optical material dispersion

In terms of the vase velocity V in a uniform medium given by

$$v = \frac{C}{i}$$

Where C is the speed of light in a vacuum and i is the refractive index of the medium.

The most commonly observed consequence of dispersion is the separation of white light into a spectrum of colours by means of a triangular prism. From

Snell's Law we can see that the angle of refraction of light in a prism depends on the refractive index of the material from which the prism is made. Since the refractive index varies with wavelength, it follows that the angle at which light is refracted also varies with wavelength, resulting in angular colour separation, also known as angular dispersion.

For visible light, the majority of transparent material has :

Whatever the refractive index n, it decreases with increasing wavelength. In this case, the medium is said to have normal dispersion. On the other hand, if the index increases as the wavelength increases, the medium has normal dispersion.

At the interface of such a material with air or with a sticky vacuum (the index is 1), Snell's law states that light incident at an angle to the normal is refracted at an angle. Then, blue light, with a higher refractive index, will be at a steeper angle to red light, which creates the rainbow.

The dispersion of light in the glass of a prism is used to build spectrometers and spectroradiometers. They are also used in spectrographic lattices, as they allow more precise discrimination of wavelengths. Dispersion in the lenses produces chromatic aberration, an undesirable effect that can distort images in microscopes, telescopes and photographic lenses. (www.google.dispersion in optics, accessed on 9 August 2023 at 22:43)

D. Looking at coloured cardboard through a prism

In order to understand the colour phenomena associated with refraction, Newton conducted a series of experiments that would become famous, the first of which involved observing coloured cardboard through a prism. The prism is a transparent block of glass, and the two refractions that occur when light passes from the air to the glass, and then from the glass to the air, are in the same direction (unlike in the case of a parallelepiped, where the refractions cancel each other out and the incident light emerges in the same direction). He then observed that the apparent position of a red cardboard and a blue cardboard were different. The path of the light is different in the two cases, which means that the refraction of blue light is different from that of red light.

E. A historic experience

This result was confirmed by Newton's second, much more original experiment. Using a hole drilled in a violet, he let a fine brush of light enter the room containing his experiments and passed this brush through a prism. He then observed that the light coming out of the prism spread out in a multitude of coloured beams, reproducing the colours of the rainbow.

The appearance of colours through a prism had already been observed before Newton. Newton's great contribution came from the following experiment, sometimes called the "experimentum grucis", which means "key experiment".

It involves passing some of the light dispersed by the first prism through a second prism. Newton thus showed that the colour was not altered by the passage through the second prism.

Newton conducted a large number of other variations of experiments, presented in his book "Opticks". In particular, he showed that recombining these coloured beams produced a beam of white light.

F. First experience

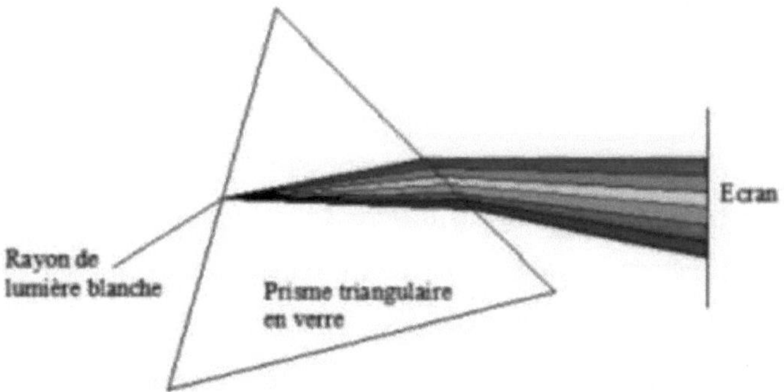

A beam of white light is passed through a glass prism and a screen is placed in front of the refracted rays. The result is a rainbow-like display of colour. This phenomenon is called light dispersion by a prism.

G. Second experiment

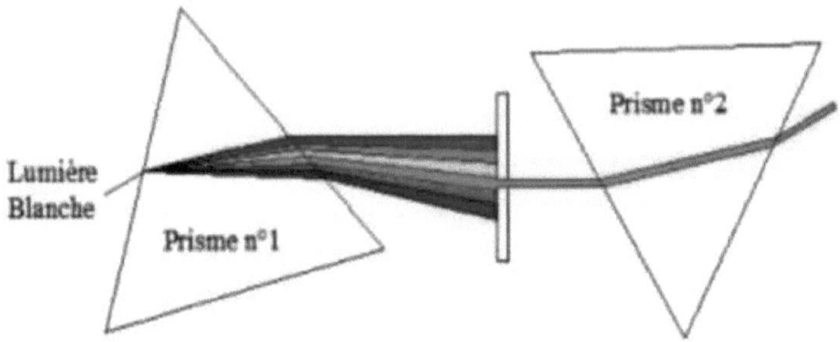

We carry out the same experiment as the first, and we capture a ray of one of the wavelengths through a screen with a hole in it, which we pass through a second prism. We can see that this beam of wavelength just undergoes a double refraction; in fact this ray does not undergo the phenomenon of dispersion because it is composed of a single wavelength, unlike white light which is composed of an infinite number of wavelengths.

H. Interpretation of results

Newton interprets it as follows:

White light is made up of rays associated with different colours, corresponding to different refractive indices. From this point of view, colours are a physical property of light (we now know that the concept of colours is more complex). The fact that the refractive index is different for different types of light is now called "dispersion". Newton couldn't really determine the refractive index of light.

the physical property of light that causes a ray to correspond to one colour rather than another.

The discovery of the phenomenon of dispersion enabled Newton to provide the first scientific explanation for the rainbow phenomenon, a phenomenon that in the previous experiment the prism was replaced by drops of water (www.google,dispersion de la lumière par le prisme consulted on 6 August 2023 at 20:03).

I. Note

The light emitted by sleep or a lamp is called "white light", which is the superposition of all the colours.

Can all the lights be decomposed?

1.1.3. Experiment with light emitted by a laser

A. **Laser**

The word laser means "amplification of light by stimulated emission of radiation" (prof GEORGES KABAMBA MWENDA KAZADI, ***questions spéciales de physique***, éd 2019, ISP/MBM).

B. **Experience**

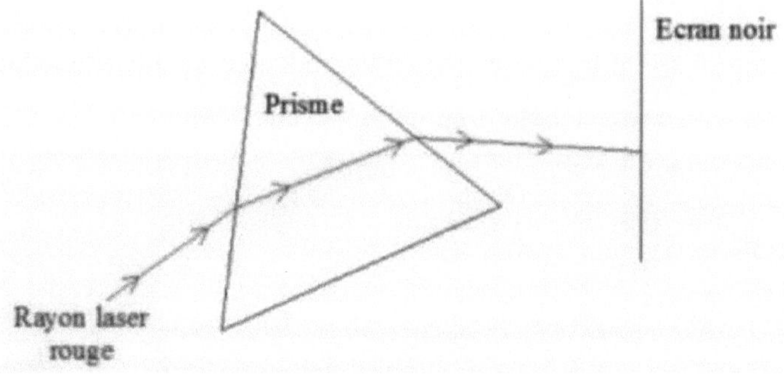

C. Comment

C. A fine red line

D. The laser beam is deflected and the spectrum contains only one colour, the initial colour of the beam emitted.

1.1.4. Wavelength

A. Definition

The wavelength is a physical quantity characteristic of a monochromatic wave in a homogeneous medium, defined as the distance separating two consecutive maxima of amplification.

B. Wavelength unit

For light, which is an electromagnetic wave visible to the human eye, the wavelength is expressed in nanometres (nm).

C. Write-off

Monochromatic light is called chromatic radiation.

All monochromatic radiation has an associated wavelength in a vacuum, denoted /. It is expressed in metres, or more generally in nanometres or micrometres.

D. Example

The red monochromatic light emitted by a laser is radiation of wavelength λ = 632.*8nm* in a vacuum.

White light is made up of an infinite number of monochromatic radiations.

1.1.5. Different wavelength ranges

A. Visible range

The spectrum of white light contains all the radiation visible to the human eye, i.e. radiation with a wavelength Λ between 400 and 700nm.

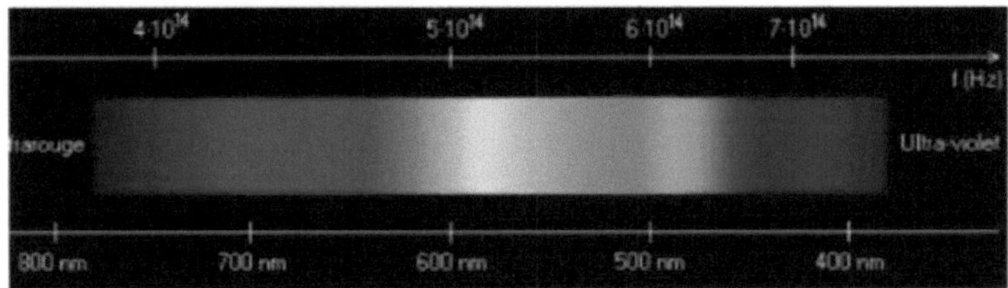

B. Other write-offs

The light spectrum extends beyond red and violet: white light contains radiation that is invisible to the human eye.

C. The electromagnetic spectrum

S When white light passes through the prism, we see a rainbow-like display of colour. The prism is said to decompose white light.

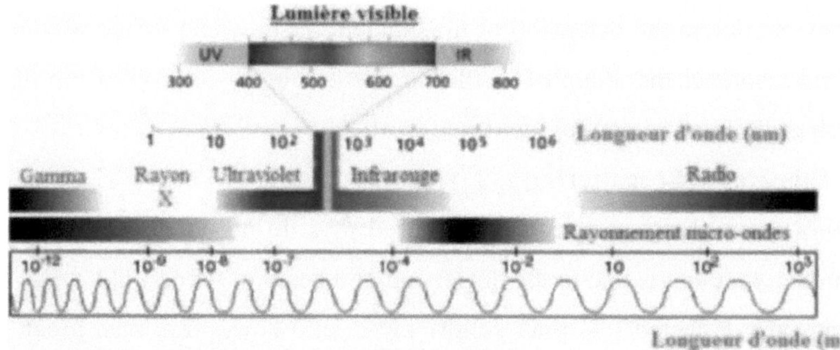

2.4.LIGHT SCATTERING BY NETWORKS

2.4.1. Dispersive network

A dispersive grating consists of a transparent support on which a large number of fine, parallel and equidistant lines have been engraved.

Each line represents a slit that diffracts the light.

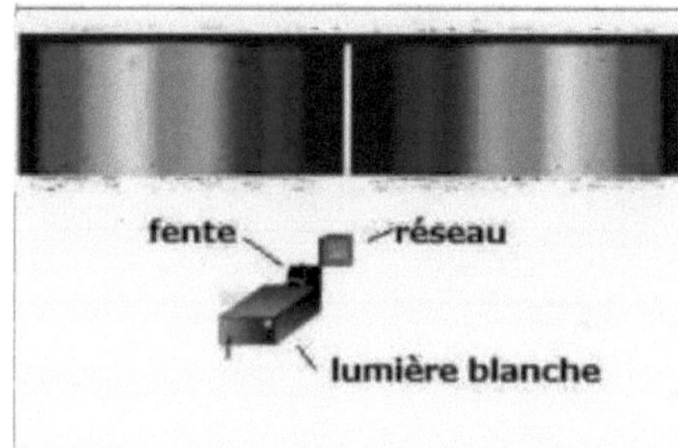

2.4.2. Diffraction of monochromatic light by a grating.

The grating is illuminated with monochromatic light. The light passing through the grating is diffracted by the slits. The interference between the different diffracted beams gives maxima of light in certain directions characterised by the angle ". κ

The angles $\ddot{\imath}$. κ are given by the relation: sin $\ddot{\imath}$ - sin ì'k = $k\imath l\lambda$

I is the angle of incidence, $\ddot{\imath}'k$ is the angle of emergence or diffraction, k is the order of the spectrum, n is the number of vtraits per metre and λ is the wavelength expressed in metres.

In monochromatic light, we obtain very fine hits, parallel to the grating slits and corresponding to different values of the order k.

For k=0, we obtain the extension of the incident beam whatever the wavelength used and whatever the angle of incidence, there is no dispersion of the light in the case of illumination under incidence, $i = 0$

2.4.3. Cause of light scattering

Newton's experiments showed that coloured rays do not all undergo the same deviation, a fact that is due to the dispersion of a beam of white light.

Red rays are deflected less than yellow rays, yellow rays less than green rays, etc. Violet rays undergo the greatest deflection.

On the other hand, the dispersive power is not the same for all transparent media. The dispersive power of carbon sulphide, for example, is higher than that of water. The table below shows that the difference in refraction between red and blue rays is 0.69 for carbon sulphide, whereas it is only 0.012 for water.

BY ADDITIVE MEANS, THE REFRACTIVE INDICES

Beam colour	Red	Yellow	Green	Blue
Carbon sulphide	1,610	1,629	1,654	1,479
	1,329	1,333	1,338	1,341

2.4.4. Mixed colours

There are two ways of mixing colours.

1) By additive process

Using three projectors of equal intensity, let's project the three fundamental colours onto a white screen. We will see that mixing these colours gives rise to the following hues:

> Blue-green gives blue-green
> Blue red gives purple (magenta)
> Green red gives yellow

Together, these colours provide white.

2) Subtractive method

If part of the rays of the spectrum are subtracted by means of a coloured filter, the remaining rays produce a specific colour (remember the experiment with potassium permanganate and potassium dichromate solutions): the colour found was violet. This colour was therefore obtained subtractively.

By adding back to the spectrum the rays that have been subtracted from it, we obtain the complete spectrum, which produces white light. The two colours that mixed to produce white light are complementary colours.

2.5.EXPERIMENT IN THE DECOMPOSITION OF LIGHT BY A PRISM

2.5.1. Equipment

- The prism
- Lamp emitting white light

2.5.2. Experience

The prism is placed in front of a small light source, and the light beam passing through the prism is broken down into all the colours of the rainbow.

2.6.RECOMPOSING WHITE LIGHT

2.6.1. What is ordinary white light made up of?

Newton, in his vision of the whiteness of sunlight resulting from all three primary colours mixed together, failed to show that the combination of just two colours (red/blue-green or yellow/blue-violet pairs) would produce white.

YOUNG (at the beginning of the 19ème century) showed that white light is not a single physical object and foresaw the tri-variance of colour perception due to the presence of three types of detectors in the retina. White light is made up of coloured lights called "radiations": blue, green and red, which are the primary colours, and yellow, cyan and magenta, which are the secondary colours.

The light spectrum is the set of coloured lights observed. There are two non-visible colours, ultraviolet and infrared. Analysis of sunlight distinguishes 7 colours: violet; indigo; blue; green; yellow; orange; red.

In fact, the spectrum of white light contains an infinite number of colours - it's a continuous spectrum.

Light is made up of radiation, which includes, in ascending order: gamma rays; X-rays; ultraviolet light; visible light (all colours); infrared light (from red light to microwaves) and radio waves.

Only colours and white light are visible to the naked eye. We note: a wavelength of 1nm at a frequency close to 3.10^{17} Hertz.

The main frequencies are :

- Gamma ray 0.1nm to 1nm
- X-ray 1nm to 100nm
- Ultraviolet 100nm to 400nm
- Visible light 400nm to 700nm
- Infrared 700nm at 1cm
- Radio wave 1cm to 1km

The particles carrying the light energy are called photons and have no mass.

COLOURS WAVELENGTHS (NM)

Extreme violet	400
Medium violet	420
Violet blue	440
Medium blue	470
Blue-green	500

Medium green	430
Green yellow	560
Medium yellow	580
Orange yellow	590
Medium orange	600
Red orange tree	610
Medium red	650
Extreme red	780

Light travels in waves (of different lengths) picked up by three series of cones in the eye. Each of these is more sensitive to one length than another. This length is then interpreted by our brain as being a "stream". (www.google, composition of light, consulted on 16 September 2023 at 169:47.

A coloured filter only transmits the colour corresponding to its own colour; it observes other colours.

The different shades of coloured light can be obtained by adding the three fundamental colours (primary colours) seen by the eye.

> Red + green + blue equals white
> Red green equal yellow
> Green + blue equals cyan
> Blue red equal magenta

2.6.2. Experimenting with the composition of white light

A. Equipment

Light sources, three blue, red and green colour filters.

B. Experience

We place a blue filter on one light source, we place the red filter on the other light source and the same for the green. The three rays intersect and become white (H. **BRASSERT and H. SOVENIER, Physique générale, Paris Dunod).**

CONCLUSION

❖ The plug breaks up white light. This phenomenon is called "white light dispersion".

- White light is made up of an infinite number of coloured lights ranging from violet to red. These coloured lights form what is known as the "white light spectrum".
- The seven main colours of the white light spectrum are: violet, indigo, blue, green, yellow, orange and red.

CHAPTER III

LIGHT DISPERSION APPLICATIONS

3.1.EXERCISES ON LIGHT

3.1.1. Exercise descriptions

1) The plate of a siren has 35 holes and rotates at a speed of 620 revolutions per minute, given that the speed of sound in air is 340 m/s.

1° the frequency emitted by the siren is ?

2° the distance travelled by the wave during one period is ?

2) Find the frequency of sound emitted by a siren consisting of a disc that makes 20 revolutions per second and has 15 holes in it.

3) The observation screen is 1.2m from the two slits and the distance between the two slits is 0.03mm. The second order bright fringe is 4.5cm from the central fringe.

a) Determine the wavelength of light

b) Calculate the distance between consecutive bright fringes.

4) A disc painted black has a white circular spot on it and rotates at a rate of 1000 revolutions per minute. It is illuminated by a stroboscope with a flash frequency of N'. Knowing only that there are three motionless white spots. What is the frequency of the flashes?

3.1.2. Resolution 1: UNKNOWN DATA FORMULAS+APPLICATIONS DIGITAL

N = 35 trous 1° F = ? $F = n.N$

N = 620 tours/min 2° λ = ? $F = 361{,}6Hz$

C = 340 m/s $\lambda = \frac{C}{F} = \frac{340}{361{,}6} = 0{,}94m$

2. UNKNOWN DATA FORMULAS+APPLICATIONS

DIGITAL

N = 15 trous	F = ?	$F = n.N$
N = 20 tours/min		$F = 15.20Hz$
		$F = 300Hz$

3. UNKNOWN DATA FORMULAS+APPLICATIONS

DIGITAL

D = 1,2m	λ = ?	$X_2 = 2.\frac{\lambda.D}{a}$
a = 0,03 mm = 0,03.10^{-3}m	f = ?	$\lambda = \frac{a.X_2}{2D}$
		$\lambda = \frac{3.4,5.10^{-7}}{2.1,2}$ m

3.2.EXERCISES ON THE REFLECTION AND REFRACTION OF LIGHT

LIGHT

3.2.1. Exercise descriptions

1) A beam of light with a wavelength of 550nm passes through a transparent plate. The incident beam forms an angle of 40° with the normal and the refracted beam forms an angle of 26° with the normal with C = 3.10^{-8} m/s. Determine the speed of light in this material and the wavelength.

2) What is the limiting angle corresponding to a medium with an index of 1.6?

3) Calculate the lateral displacement produced by a glass slide (n=j d=2cm thick) at an incidence of 30°C.

4) Circle where there are good suggestions

a. the approximation that light propagates in a straight line is valid in :

1. isotropic, transparent and homogeneous media
2. homogeneous, absorbent and isotropic media
3. the air

4. water
5. all environments.

b. in a material environment :

1. the frequency is increased
2. the frequency is reduced
3. the frequency is identical
4. the speed of light increases
5. the speed of light decreases
6. the speed of light remains the same.

c. total reflection can occur when passing from a medium of index n1 to a medium of index n2

1. if n n_1 2 and for low angles of incidence
2. if n n_1 2 and for low angles of incidence
3. if n1 n2 and for large angles of incidence
4. if ni П2 and for smaller angles of incidence

d. when refracting through an air/water dioptre, a ray coming from the air :

1. is close to normal in water
2. deviates from normal in water
3. never even partially thought through

e. light travels through water than through :

1. the void
2. the air
3. glass

5) In which century did Descartes live? What area of optics was he interested in?

6) What are the characteristic quantities of a sine wave?

7) What wavelength range does visible light correspond to? What is the wavelength of red light?

8) Define the index of a medium.

9) Recall Descartes' laws for the reflection of a ray of light. 10) A tank contains

a layer of water topped by a layer of benzine.

a. A beam of light emanating from water is incident at 30°C on the water-benzine separation surface. What will be the angle of emergence of the light beam when it leaves the benzine?

b. At what angle of incidence must it fall on the water-benzine separation surface to be totally reflected on the benzine-air separation surface?

10) What must be the radius r of a circular cork with the length h of the needle immersed in the water so that none of it comes out of the water, given that the refractive index of water compared with air is

3.2.2. Responses

1. DATA UNKNOWN FORMULA

$\lambda_0 = 550nm$ 1° $\lambda = ?$ $n = \frac{C}{V} = \frac{\lambda_0}{\lambda} = \frac{\sin\hat{\imath}}{\sin\hat{r}}$

$\hat{r} = 26°$ $1°\ \lambda = \frac{\lambda_0}{n} \Rightarrow \lambda = \frac{550.10^{-9}}{1,468} = 374,659\ nm$

$\hat{\imath} = 40°$

$\lambda = 374,7nm$

$C = 3.10^8 m/s$ $2°\ v = \frac{C}{n} = \frac{3.10^8}{1,468} = 2,044.\ 10^8 m/s$

$v = 2,044.10^8$ m/s

2. UNKNOWN DATA FORMULA

$n = 1,6$ $\hat{\imath}_L = ?$ $i_L = \arcsin\left(\frac{1}{n}\right) = \arcsin\left(\frac{1}{1,6}\right)$

$= \arcsin 0,625$

$\hat{\imath}_L = 38,68°$

2. UNKNOWN DATA FORMULA

$n = \frac{3}{2}$ $d = ?$ $d = e.\frac{sin(\hat{\imath}-r)}{cosr}$ or $n = \frac{sin\hat{\imath}}{sin\hat{r}}$

$e = 2cm$ $\hat{\imath} = 30° \Rightarrow sin\hat{r} = \frac{sin\hat{\imath}}{n}$

$$sin\hat{r} = \frac{sin30°}{\frac{3}{2}} = \frac{1}{2}.\frac{2}{3} = \frac{1}{3}$$

$$r = arcsin\frac{1}{3} = 19{,}47°$$

$$= 2.10^{-2}\frac{sin(30° - 19{,}47°)}{cos19{,}47°}$$

$$= 2.10^{-2}.0{,}194$$

$$= 0{,}388.10^{-2}m$$

$$\mathbf{d = 3{,}88mm}$$

4. a. the approximation that light propagates in a straight line is valid in :

1. **isotropic, transparent and homogeneous**

b. in a material environment :

2. **the speed of light increases**

c. total reflection can occur when passing from a medium of incidence n1 to one of incidence n2.

3. **if ni Π2 and for large angles of incidence.**

d. when refracting a ray from the air through an air/water dioptre :

4. is close to normal in water

e. light travels faster in water than in :

5. glass

5. Descartes lived in the 17th century and contributed to the development of geometrical optics.

6. I'll mention the frequency *f* and the wavelength in a vacuum. We also know that $\lambda = \frac{c}{f}$

7. The visible corresponds roughly to λ between 400 and 800nm. Red extends from about 620 to 800nm, averaging around $\lambda R = 700$n.

8. The index n of a medium is defined as the ratio of the speed of light in the medium to $n = \frac{c}{v}$ its speed in the medium.

material.

9. Reflection: the reflected ray is in the plane of incidence and the angle of refraction is equal to the angle of incidence. $\hat{r} = \hat{\imath}_1$

Refraction: the refracted ray is in the plane of incidence and the angle of refraction i is related to n srnɩ$_{l1}$ = $n_2\ sim_2$

10. a. the angle of refraction r on entering the benzine is such that: $n_1 sin\hat{\imath} = n_2 sin\hat{r} : \frac{4}{3}\sin 30° = \frac{3}{2}\sin\hat{r} \Rightarrow sin\hat{r} = \frac{4}{9}$ the incidence of refraction of the air with respect to the benzine is :

$$n' = \frac{1{,}00029}{\frac{3}{2}} \simeq \frac{2}{3}$$

The emergence angle e at the benzine outlet is such that :

$$\sin\hat{e} = \frac{\sin\hat{r}}{n'} : sin\hat{e} = \frac{\frac{4}{9}}{\frac{2}{3}} = \frac{2}{3}$$

$$\hat{e} = arcsin\frac{2}{3}$$

$$\hat{e} = 42°$$

c. For the light ray to refract by grazing the benzine-air separation surface, the angle of refraction r at the entrance to the benzine must be such that

$$sin\hat{e} = \frac{sin\hat{r}}{\frac{2}{3}} = \sin\ 90°$$

i.e. .≡ - - the angle

3 3

of incidence ï on the water-benzine separation surface must be such that :

$$n_1 sin\hat{\imath} = n_2 sin\hat{r}: \frac{4}{3} sin\hat{\imath} = \frac{3}{2} \cdot \frac{2}{3}$$

$$sin\hat{\imath} = \frac{3}{4}; \hat{\imath} = 48°$$

The refractive index of water in relation to air is $\frac{4}{3}$, so : $sin\hat{\imath} = \frac{1}{n}$;

$sin\hat{\imath} = \frac{3}{4}$; $\hat{\imath} = arcsin\frac{3}{4}$; $\hat{\imath} = 48°$ for no rays to emerge from the water, it is necessary that

$tg\hat{\imath} = \frac{r}{h}$ d'où $r = htag48°$ $r = 1{,}11h$

3.3.EXERCISES ON LIGHT DISPERSION THROUGH A PRISM

3.3.1. Exercise descriptions

1. Indicate the î, ¿', r, r'and *D* values for a prism in the following cases:

a. Grazing incidence *(= W)*

b. 45° incidence

c. No impact

d. Grazing angle of 90° *(i* = 90)°

e. Emergence of 45° Q = 45°)

f. Minimum deviation.

2. A monochromatic ray of light passes through a glass prism with an index = 1.6 and an angle A = 30°. The incident ray falls on the prism at an angle I = 30 ". Determine the angle of refraction ?- on the first face, the angle of incidence r ' on the second face, the angle of emergence *ï'* and the total deflection created by this prism.

3. A prism of angle A and index = 1.5 is illuminated by an incident ray perpendicular to the entrance face of the prism traced by the path of the light ray and calculate the deflection *D* in the following two cases: *A = 30° and A = 60°.*

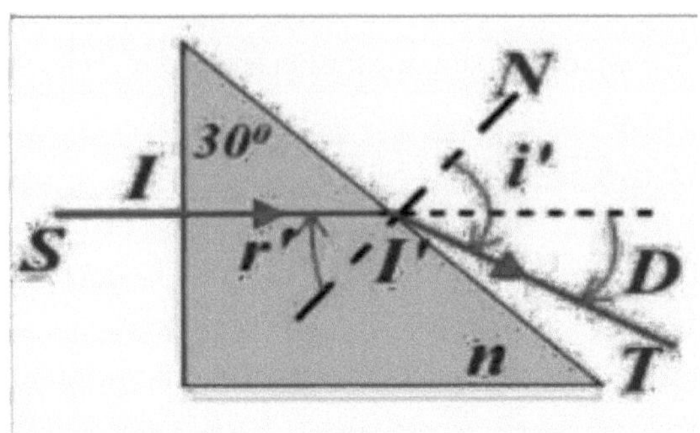

4. A prism with index = 1.5 has the cross-section of an equilateral triangle.

a. Determine the minimum angle of deflection when the prism is placed in air.

b. What is the value of the minimum angle of deflection Dm when the prism is immersed in water of index $\frac{4}{3}$.

3.4.EXRCISES ON LIGHT DISPERSION BY THE NETWORK

3.4.1. Exercise descriptions

1. A beam of monochromatic light produced by a helium-neon laser (= 632.8) has no normal incidence on a diffraction grating containing 6000 lines/cm. Determine the angles for which the first-order maximum, second-order maximum, etc. are observed.

2. A neon light is viewed through a slit 0.3mm long. At what distance should the fringes of two eyes be placed to have a gap of 1mm between the first and second dark fringes, framing the central fringe for$_\lambda$ =$_5$ OOOΛ.

3. The concentration of an aqueous solution of potassium permanganate ($^+$ *(aq)* + M_n ***O*** *4* - *(a q))* is to be determined by UV-visible spectroscopy.

A laboratory has a high-quality spectrophotometer with a monochromator that can be used in the ultraviolet, visible and infrared ranges. It includes a set of several gratings depending on the working wavelength. For this assay, an etched grating with 500 lines/millimetre is chosen.

The grating is illuminated under incidence by wavelength relations between λ_1= *400nm e t* λ_2= 800 n *m*

a) Determine the pitch of the network used

b) The fundamental formula for flat networks is $sin\theta = sini + k\frac{\lambda}{a}$

Using the diagram below, explain the meaning of the various terms in this formula, specifying the units.

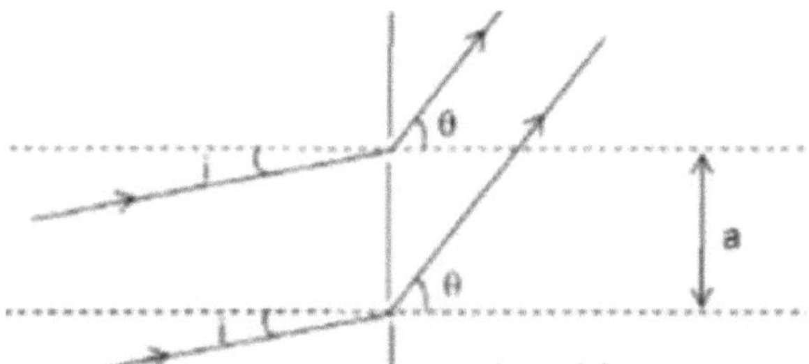

The diagram has been reworked with no concern for ecnelte

c) Calculate, in the first-order spectrum, the angles θ_1 e t θ_2 for radiation of wavelengths λ_1 e t λ_2. The grating being illuminated under normal incidence. Deduce the angular difference between the two radiations.

3.4.2. Answers

1. λ = 632.8nm

= 6000

First we need to calculate the network pitch

$$a = \frac{1}{n} \Rightarrow a = \frac{1}{6000} = 1667nm$$

For the first-order maximum (1) we obtain :

$$sin\theta_1 = \frac{\lambda}{a} = \frac{632{,}8nm}{1667nm} = 0{,}3796$$

$$\boldsymbol{\theta_1 = 2}$$

Likewise,

$K = 2, on\ a$:

$$\sin\theta_2 = 2.\frac{\lambda}{a} = 2.\frac{632{,}8nm}{1667nm} = 0{,}7592$$

$$\boldsymbol{\theta_2} = 4$$

However, for K = 3, we obtain *if* $n\theta_3$ = 1, 139. Since *if* $n\theta$ cannot be greater than one, this solution is not valid. So we only have the first two orders in this situation.

CONCLUSION

Here we are at the end of our work. Light is omnipresent in our lives. It is thanks to light that life is possible on our planet. Life could not have developed without sunlight. Even today, planets and animals need light to survive.

Light is also our main means of discovering the world around us. The rainbow is a luminous phenomenon that can be seen in the sky when the sun shines through the rain, showing different colours spread out like a ribbon in the shape of a bow. The same observation is made when light passes through a prism and a grating, which are optical systems, and we see a display of colours similar to that of the rainbow. This observation allows us to understand that the prism and the grating decompose ordinary white light, and in physics this phenomenon is called "the dispersion of white light".

We thought we'd put together a collection of experiments on light dispersion. To achieve this, we used the documentary method. On the one hand, this work will help teachers of geometrical optics to improve the quality of this branch from an experimental point of view, and on the other, it will provide learners with experiments on the dispersion of light.

Without claiming that this work is perfect, we remain open to all pertinent comments and suggestions from our readers with a view to improving it in the future. Once again, thank you to all those who helped us in the preparation of this work.

BIBLIOGRAPHY

1. Cessac, J et Treherne, G, physique classe de seconde C, Fernand Nathan, paris, 1966

2. A. Delaruelle, A.I, Claies, cours de physique, chaleur, optique géométrique, Wesmaiel Charlier (SA), Namur 1971.

3. A. Delaruelle, A.I, Claies, element of physics, heat, acoustics, geometrical optics.
Namur 1964.

4. Faucher, R, physique classe de 1èreC/D/E/, collection "Hatier", librairie Hatier, Paris 1966.

5. BRASSEUR.H, SAUVENIER.H, physique générale, Duno 1966.

6. Raymond, A, Serway, physique III, optique et physique moderne 3ème éd ; traducteur Robert Maorin et Céline Trembley éd. Etudes vivantes 1992.

7. National Physics Programme 2005.

8. HECHT, E., (2003), physique De Boeck Université, Brussels

9. J. Médard NSUNGULA TSHIBAKA, cours d'optique géométrique, application à la photographie, Ed 2005.

10. Programme éducatif du domaine d'apprentissage des sciences, 1ère edition Kinshasa 2021.

11. Professeur Georges KABAMBA MWENDA KAZADI, questions spéciales de physique 1ère licence math-physique, éd 2020-2021, ISP/MJM.

12. CT Médard NSUNGULA TSHIBAKA, notes de cours de physique 2ème graduat, éd 2019-2020, ISP/MJM.

13. Patrick KAMALENGA NGOYI, Classical studies of optical instruments, TFC, ed 2014-2015, ISP/MJM.

Printed by Books on Demand GmbH, Norderstedt / Germany